TIDE POOLS

THE SEA

Jason Cooper

The Rourke Corporation, Inc.
Vero Beach, Florida 32964

Edited by Sandra A. Robinson

PHOTO CREDITS
All photos © Lynn M. Stone

LIBRARY OF CONGRESS

Library of Congress Cataloging-in-Publication Data

Cooper, Jason, 1942-
Tide pools / by Jason Cooper.
p. cm. — (Discovery library of the sea)
Includes index.
Summary: Describes how tide pools are formed, where they are found, and some of the plants and animals that live in them.
ISBN 0-86593-234-4
1. Tide pool ecology—Juvenile literature. 2. Tide pools—Juvenile literature. [1. Tide pools. 2. Tide pool ecology. 3. Ecology.] I. Title. II. Series: Cooper, Jason, 1942- Discovery library of the sea.
QH541.5.S35C66 1992
574.5'2636—dc20 92-16074
CIP
AC

Printed in the USA

TABLE OF CONTENTS

TIDE POOLS

The ocean rises and falls each day along the seashore. Ground that a person can walk on at lunchtime will be under water by suppertime. This changing level of the sea is called the **tide.**

While the tide is high, the sea reaches several feet above its level at low tide. When the tide shifts to low, the sea level falls. But some of the seawater remains behind, trapped in rocky holes and other low places. These pools of water that remain on shore during low tides are called tide pools.

Low tide leaves pools of seawater on a rocky Pacific shore

POOLS IN THE ROCKS

Most tide pools form along rocky seashores. Bowl-shaped rocks and rocks with pits in them hold seawater because it has no way to drain.

Marine, or sea animals can safely take up housekeeping in rocky pools. During very low tides, some of the animals may be exposed to air. Most of them can live a short time out of water. Before long, the tide rises, and the pools refill with water and disappear under the sea.

Starfish and sea anemones waiting for high tide

LIFE IN THE TIDE POOLS

Tide pools are **habitats,** or homes, for hundreds of kinds of curious and colorful marine creatures. Seaweeds, plants in the **algae** family, also live in the tide pools, giving many of them a greenish color.

No two tide pools are alike. How deep a tide pool is and how near it is to the edge of the sea affect the kinds of animals and algae that live in it.

The most interesting tide pools are rich with such animals as mussels, starfish, sea anemones, crabs, worms and sea urchins.

Barnacles, mussels, sea anemones, starfish and algae share an Oregon tide pool

THE TIDE POOL SHORES

Tide pools are found wherever seashores are rocky. On the Pacific coast of North America, the seashores of California, Oregon, Washington and British Columbia are favorites of tide-poolers—people who explore tide pools.

On the Atlantic coast, Nova Scotia, New Brunswick and Maine have rocky shores and plentiful tide pools.

Tide pools do not last long on sandy beaches. Sand shifted around by high tides quickly fills them.

Rocky Pacific coast of central Oregon is a tide pool wonderland

Purple sea urchins and giant green sea anemones in an Oregon tide pool

Crab, green sea urchins and snails in a Maine tide pool

ARMED ANIMALS

Among the largest and most remarkable of the tide pool animals is the starfish, or sea star. It has five or more replaceable "arms." If a starfish loses one of its arms, a new one grows in its place.

The starfish is one of the tide pool **predators.** Predators are hunters that feed, or **prey,** on other animals.

The starfish preys upon such creatures as the mussel clams. The tiny, sucking feet under each starfish "arm" can pull a mussel shell open.

Atlantic starfish feeding on mussels in an Atlantic tide pool

ANIMALS OR FLOWERS?

Sea anemones brighten every tide pool in which they are found. Sea anemones are named for the anemone flower. But sea anemones are animals. What appear to be flower petals are actually the sea anemone's **tentacles.**

The tentacles ring the sea anemone's mouth. They release a poison that helps the anemone kill the little fish that are its prey. The tentacles are harmless to people.

Flowerlike green sea anemones display their tentacles

SHELLS AND SPINES

Tide pool rocks are often crusty with little hard-shelled creatures called barnacles. Barnacles are relatives of the crabs that also live in tide pools.

Many tide pool animals are **mollusks.** The mollusk family includes octopuses and the strange, little chiton with its suit of shell armor. Most mollusks in a tide pool are snails and mussels.

The sea urchins in a tide pool may be red, black or purple. Sea urchins wear an armor of spines that protect them from many predators.

Barnacles (white) and chiton cling to rocks on a Pacific shore

EXPLORING TIDE POOLS

A rocky tide pool is a zoo and a garden. And each of its dark, little corners and tangles of seaweed is a home to marine animals.

A tide pool may be one of the most amazing places you ever explore. But approach it carefully. Seaside rocks are extremely slippery, and tide pools may be deep. Sea foam and seaweed sometimes hide tide pools altogether.

Always explore the rocky shores with someone else, and keep an eye out for incoming waves.

Tide-poolers explore Oregon's Marine Gardens Shore Preserve during low tide

PROTECTING TIDE POOLS

You can explore the world of the tide pool by looking carefully into it, or by carefully reaching into it.

Never pull tide pool animals from the rocks, and never take these marine creatures home. They will not live long if they are removed from their natural habitat.

Some states wisely protect tide pools by law. Oregon, for example, has many excellent seashore **preserves,** or protected places, for its marine animals.

Glossary

algae (AL jee) — a group of nonflowering plants, many of which are seaweeds and live in salt water

habitat (HAB uh tat) — the kind of place in which an animal lives, such as a tide pool

marine (muh REEN) — of or relating to the sea, salt water

mollusk (MAHL usk) — a group of soft, boneless animals usually protected by hard shells of their own making; for example, clams, oysters, snails

predator (PRED uh tor) — an animal that kills other animals for food

preserve (pre ZERV) — an area where wild animals and plants are protected from harm by people

prey (PRAY) — an animal that is hunted by another for food

tentacles (TEN tah kulz) — a group of long, flexible body parts usually growing around an animal's mouth and used for touching, grasping or stinging

tide (TIDE) — the daily rise and fall in the level of the oceans and seas

INDEX